THE A B C OF SOUTHERN ELECTRICS

Up from the sunshine coast ! A Littlehampton express approaching the Ouse Valley Viaduct

Compiled and published
by
IAN ALLAN
225-7 Laleham Road
Staines

Part I

GENERALLY speaking, Southern electric multiple unit trains can be divided into two classes—suburban and main line stock.

First let us deal with the suburban sets which are for the most part made up of 3-car units having two 275 horse-power motors in each of the motor coaches, which are situated at either end of the unit; between each two motor coaches is one trailer car. For rush hour workings, two of these 3-car sets are employed with a two coach trailer set in between.

No definite classification can be given to the various units as there are many varieties—some are rebuilt from L.S.W.R., some from L.B.S.C.R. and some from S.E.C.R. steam stock, while quite a large proportion has been built to new design.

Another type of suburban unit is that employed on the South London line between Victoria and London Bridge and the West Croydon/Wimbledon service. These are in two coach sets and the former have been rebuilt from the old "Brighton overheads." Their serial numbers are 1801–1812.

A new arrangement of suburban stock is just appearing on the Western Section where 3-car L.S.W.R. and L.B.S.C.R. units have been remarshalled with trailer cars into 4-car sets (motor coach at each end, two trailers in between). This does away with the 2-car trailer sets between the units for peak periods mentioned above.

* * *

There are several types of express main line stock which include 4- and 6-car corridor sets, some having Pullman, Restaurant and Buffet cars included. In May, 1942, however, all these cars were withdrawn for the duration of the war (except the restaurant cars, which are still running, though without service), and have been taken out of the units, so that a set formerly having six coaches, including a Pullman car, and previously coded "6 Pul.," will now only have five cars and be coded "5 Cor."

On page 10 will be seen diagram plans of several of these types and below are full descriptions.

An 8-car suburban train near Streatham Common. The leading 3-car set is of ex-S.E.C.R. steam stock

6-car Express Corridor units (with Pullman car).—For express electric services between London and Brighton, Littlehampton and Hastings, 6-car corridor units with Pullman cars were in regular use prior to 1942.

The motor coaches of these units embody the best practices of the carriage builders. The interiors of the saloons are finished in mahogany with figured panelling and above the cornices the ceilings are painted with white enamel. Transverse seats, divided by a central gangway, give accommodation for two passengers in each seat, and above the seats net racks are provided. Table fixings are also fitted in the coaches for the use of a limited number of portable tables, which are stored in the guard's compartment.

Motor coach of a 6-car Pullman unit

With the object of improving passenger movement between the coaches, the motors and trailers are fitted with a flexible communication gangway of greater width than the British standard type. The equipment of the guard's and luggage compartment includes periscope, seat, desk, letter rack, brake valve, expanding ladder, shoe paddle and emergency equipment.

It had been the practice of the company to build its motor bogies for suburban traffic on the lines of standard British type carriage bogies, but it was decided, after extensive trials, to fit the new express stock with the equalising beam type. The position of the collector shoe and shoe beam has, however, necessitated some departure from the orthodox design of this type of bogie.

In the trailer coaches, no departure from the Company's standard body construction has been made. The bodies have timber frames encased in metal panelling at the sides and ends, with deal roofing covered with canvas. The floors are, between the cross bearers, fitted with dove-tail sheeting and cork slabs in a manner similar to the motor coaches. The underframes are steel.

The electrical equipment of the train units has been designed to give a maximum service speed of 75 m.p.h. Each motor coach is equipped with four 225 h.p. motors and one set of electro-pneumatic control gear arranged for series-parallel control.

For the "City Limited" service three 6-car Pullman sets were provided, made up of two 3rd motors, three 1st class corridor trailers and a composite Pullman car. These are coded 6-Cit.

6-car Express Corridor units (with Pantry car).—The 6-car "Pan" units differ from the Pullman stock by the substitution of a refreshment car for the Pullman kitchen-parlour car. The layout of the train apart from this is practically the same, but the formation consists of 2 motor brake thirds, 2 corridor thirds and 2 corridor firsts, one of which includes a pantry portion.

The first-class pantry car is a beautifully upholstered vehicle containing 5 first class compartments with removable tables. The kitchen is at one end of the vehicle and has two serving hatches. Heating of the passenger compartments in all the trailers is effected by a tubular electric heater fitted beneath each seat; the heaters are fed from separate circuits so arranged that one or both may be on as required. In the open saloons of the motor thirds, the heaters are of the spiral-wound open-coil type.

At busy times when 12-car trains are required one Pullman unit and one pantry unit are coupled together so that refreshment facilities are available to all passengers, although there is no corridor communication throughout the train.

4-car Express units.—The latest type of S.R. express electric stock is the 4-car continuous corridor unit—now in regular service on the Portsmouth (both Central and Western Sections) and Bognor lines.

The motor coaches in the 4-car sets differ from the Brighton and Eastbourne vehicles (6-car corridor sets) in being carried on one

An express composed of a 6-car Pantry unit (6-pan)

An 8-car Portsmouth express made up of a 4-car corridor unit and a restaurant/kitchen car set

motor bogie and one trailing bogie. There are two motor coaches with an aggregate of four motors in each 4-car unit (two motors to each motor bogie). The motor bogie is of the same type as that fitted to the Brighton stock and the trailing bogie is of the normal suburban stock design.

Another new feature of the 4-car sets is that vestibuled connections are provided at both ends of all 4-car trains, so that whatever the formation there is end-to-end communication. The passage through the vestibules, past the driving cabin, at each end of the 4-car set has been arranged in an ingenious manner, so as completely to cut off the motorman's compartment when not in use.

The formation of the ordinary sets can be seen in the diagram on page 10, but some units contain restaurant or buffet cars. In the restaurant-kitchen units the first class restaurant car is a composite dining saloon and ordinary first class compartment vehicle; while the third class restaurant car is a combined diner and kitchen car.

An all-electric kitchen with most comprehensive equipment is situated between the first and third dining saloons.

SOME MAIN LINE STOCK INTERIORS

Open type third class motor coach

The bar section of a buffet car

Restaurant car, with tables laid for lunch

BUFFET CARS

The buffet cars, which are included in some units, are a new type of vehicle so far as the Southern Railway is concerned. Each car contains a kitchen, a bar section, a saloon section and two lavatories separated from the saloon section by a cross gangway. The saloon section had to provide seating for 16 passengers and the problem was solved in an original manner by fitting specially shaped tables facing each window. The edge of the table top consists of four concave curves at each of which a revolving chair is provided. This arrangement gives each passenger ample table room and, at the same time, provides a greater circulating area and thereby facilitates the service. The front edge of the bar counter also consists of a series of curves, and a high stool with a circular top, free to revolve, is placed at each. The edge of the intermediate partition has been made to follow the same design, which has also been applied as a decorative motif to the end partition. Special attention has been paid to silencing the cars. Double glass windows are fitted, the undersides of the floors have been covered with sprayed asbestos, and the walls and roofs have been lined with acoustic asbestos blankets.

* * *

For semi-fast and slow services 2- and 4-car non-vestibuled corridor and non-corridor coaches are employed. Here follow full details of each type.

4-car Corridor/Non-Corridor Lavatory units.—The design of this stock was materially influenced by the longer journeys for which these units were to be used as compared with the Company's suburban electric rolling stock, and special attention was given to the seats and upholstery; whilst the space afforded to each passenger is in keeping with the newest main line standards.

A 4-car unit near Three Bridges, working a Victoria-Brighton stopping train

Each four-coach unit provides seating accommodation for 70 first class passengers distributed as shown in the diagram on page 10. Thirty-three of these four-coach units were constructed for the extension to Brighton and Worthing in 1932, each consisting of two motor coaches and two trailers. The bogie under the driver's compartment is fitted with two 275 h.p. motors, the power of a unit therefore being 1,100 h.p.

2-car Non-Vestibule Corridor units.—The 2-car non-vestibule corridor sets are for use on semi-fast and slow trains on the Central Section and on Waterloo—Reading, Alton and Portsmouth (slow) services on the Western Section. One of the vehicles is a motor third and the other a composite driving trailer. Lavatory accommodation is provided in each. A guard's and luggage compartment, with approximately 60 sq. ft. of floor space is provided in the motor coach. The total seating capacity is 24 first class and 84 third class, and the tare weight of the set is 75 tons. This stock is 9 ft. wide.

2-car Corridor and Non-corridor Non-Vestibuled units.—For the extension to Gillingham and Maidstone a total of 76 2-car " Hal. " units were built and, in view of the outer-suburban rather than suburban character of the services, a corridor has been introduced in half of the stock. Each twin-car set comprises a motor third of compartment type and a composite driving trailer with side corridor and lavatory accommodation, but without vestibule connexions. The right type of accommodation is thus provided both for the short and long-distance passenger. The electrical equipment comprises two of the standard nose-suspended motors, with an individual one-hour rating of 275 h.p. mounted on the outer bogie of the motor coach.

2-car non-Corridor units.—The 2-car non-corridor units are formed of one motor third brake and one driving trailer composite permanently coupled together. The dimensions and appearance of these vehicles are generally similar to those of the standard 8 ft. wide suburban stock, but some motor coaches include a third class coupé compartment, a feature found in certain trains running in the London district. The electrical equipment of these sets corresponds generally with that on suburban trains, but only two 275 h.p. self-ventilated nose-suspended motors are used for the 2-car set taring 71 tons, compared with the four on a 3-car suburban set taring 101 tons ; but, of course, with an 8-car train, the same number of motors is provided in each type of stock, with 4 motor coaches and 4 trailers.

2-car (Bil.) Corridor unit [*Photo: "Railway Gazette"*

2-car (Hal.) Corridor/Non-Corridor unit

2-car (Nol.) Non-Corridor unit

EXPRESS STOCK

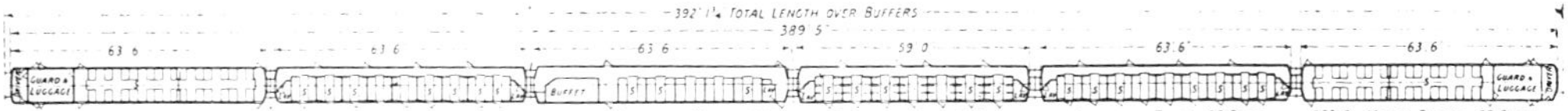

6-Car Motor Corridor Units—with Pantry Car (6-Pan.)

6-Car Motor Corridor Units—with Pullman Car (6-Pul.)

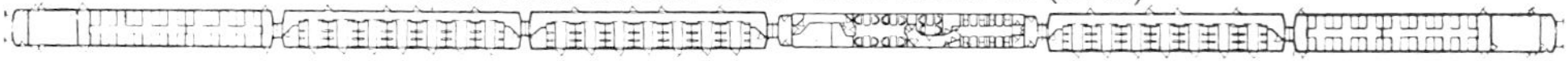

6-Car " City Ltd " Units

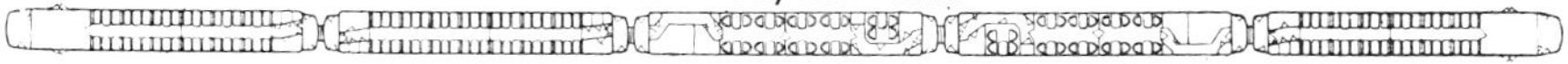

5-Car Brighton Belle Units

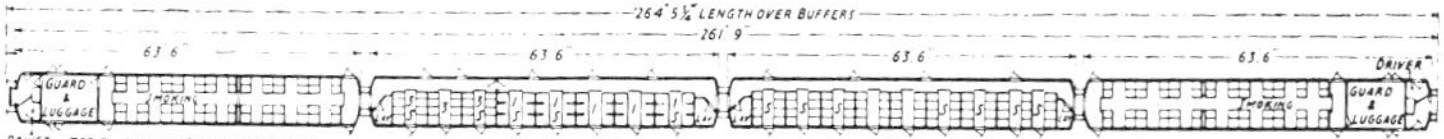

4-Car Motor Corridor Units (4-Cor.)

SEMI FAST AND SLOW STOCK

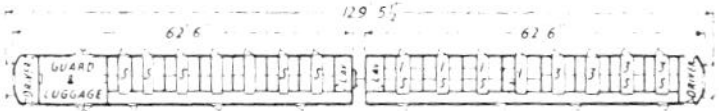

2-Car Motor Corridor Units (2-Bil.)

4-Car Motor Lavatory Units (4-Lav.)

WATERLOO & CITY LINE STOCK

Not, perhaps, naturally associated with the Southern Electric system, but none the less an important part of it, is the Waterloo and City line. Recently, completely new stock was provided and the whole railway modernised. Twenty-eight new coaches were supplied—12 motor and 16 trailer—which are formed into five 5-car units (two motors and three trailers in each) with three spare vehicles.

During peak hours 5-car trains, accommodating 600 passengers are run, while during slack periods the service is maintained by single motor coaches, detached from the 5-car sets.

The control equipment of the trains is of the most modern design, and is similar to, and largely interchangeable with, the equipments used on the latest suburban and express units.

The coaches are provided with electrically controlled air-operated doors, so that the guard is able to control all the doors on a train from one position. Two double doors per side are provided on each trailer coach, giving ample opening for the rapid loading and unloading of passengers. There is a door control in each driving cab and telephone communication is provided between the motorman and guard by means of Loudaphones fitted in each driving cab.

The Waterloo and City Railway is London's shortest tube railway, being only 1 mile and 46 chains long, and having but two stations.

ELECTRIC LOCOMOTIVES

The Southern has three electric locomotives in service. The largest is the now famous CC 1 mighty 100-ton loco, capable of hauling passenger trains up to 75 m.p.h. or freight trains up to 1,000 tons.

The locomotive consists of a box type cab carried on two six-wheel bogie trucks. Each of the six axles is driven by an electric motor through single reduction spur gearing.

The box cab contains an electrically fired boiler for supplying steam for heating passenger trains as well as electrically-driven compressors and exhausters for operating the brakes—it also contains the electrical apparatus for controlling the traction motors.

At each end there is a motorman's driving cab containing his controller and the brake operating levers, together with a speedometer and all the necessary instruments associated with the electrical braking and steam-heating equipment.

Current is collected from the conductor rail by eight shoes mounted on each side of the locomotive, and from overhead wires

Electric Locomotive CCI

(Courtesy S.R. Magazine)

(which have been installed in certain sidings) by means of a pantograph.

The controller has 26 positions or notches for controlling the voltage applied to the traction motors. Each position is a " running " position which can be used for any length of time, thus giving very smooth acceleration and a large number of " running " speeds.

CC 1 is so designed that it can continue to exert a drawbar pull when passing over the unavoidable " gaps " in the conductor rail which occur at junctions and crossings.

The other two locos are much smaller and are both service engines. No. 74S, which is mounted on two 4-wheel bogies, was built to Mr. Drummond's design, and started its career on the Waterloo and City Railway ; but, with the introduction of motor bogie stock, came above ground and is now to be seen at Durnsford Road, where it is employed on the haulage of coal wagons.

The smallest of the electric locos is the 4-wheeled 75 S, still at work on the Waterloo and City line—used for moving " dead " passenger vehicles and for hauling coal-trucks.

Both of the latter locos take current from the third rail in the usual way.

Part II

Numerical list of electric motor unit numbers

(to be seen on front of each set)

3-Car Suburban Sets

1203	1273	1410	1452	1498	1585	1628	1670
1205	1274	1411	1453	1499	1586	1629	1671
1207	1276	1412	1455	1500	1587	1630	1672
1208	1277	1413	1456	1501	1588	1631	1673
1209	1278	1414	1457	1502	1589	1632	1674
1211	1282	1415	1458	1503	1590	1633	1675
1214	1284	1416	1459	1504	1591	1634	1676
1215	1285	1417	1460	1505	1592	1635	1677
1216	1286	1418	1461	1506	1593	1636	1678
1217	1287	1419	1463	1507	1594	1637	1679
1218	1288	1420	1464	1508	1595	1638	1680
1219	1289	1421	1465	1509	1596	1639	1681
1220	1290	1422	1466	1510	1597	1640	1682
1221	1291	1423	1467	1511	1598	1641	1683
1222	1292	1424	1468	1512	1599	1642	1684
1223	1293	1425	1469	1513	1600	1643	1685
1224	1294	1426	1470	1514	1601	1644	1686
1225	1295	1427	1471	1515	1602	1645	1687
1226	1296	1428	1472	1516	1603	1646	1688
1228	1297	1429	1474	1517	1604	1647	1689
1230	1298	1430	1475	1518	1605	1648	1690
1233	1299	1431	1477	1519	1606	1649	1691
1236	1300	1432	1478	1520	1607	1650	1692
1238	1301	1433	1479	1521	1608	1651	1693
1241	1302	1434	1480	1522	1609	1652	1694
1242	1303	1435	1481	1523	1610	1653	1695
1243	1304	1436	1482	1524	1611	1654	1696
1244	1305	1437	1483	1525	1613	1655	1697
1245	1306	1438	1484	1526	1614	1656	1698
1249	1307	1439	1485	1527	1615	1657	1699
1250	1308	1440	1486	1528	1616	1658	1700
1251	1309	1441	1487	1529	1617	1659	1701
1252	1310	1442	1488	1531	1618	1660	1702
1253	1401	1443	1489	1532	1619	1661	1703
1256	1402	1444	1490	1533	1620	1662	1704
1261	1403	1445	1491	1534	1621	1663	1705
1263	1404	1446	1492	1579	1622	1664	1706
1265	1405	1447	1493	1580	1623	1665	1707
1266	1406	1448	1494	1581	1624	1666	1708
1267	1407	1449	1495	1582	1625	1667	1710
1268	1408	1450	1496	1583	1626	1668	1711
1270	1409	1451	1497	1584	1627	1669	1712

3-Car Suburban Sets—*continued.*

1713	1725	1736	1749	1761	1772	1782	1792
1714	1726	1737	1750	1762	1773	1783	1793
1715	1727	1738	1751	1763	1774	1784	1794
1716	1728	1739	1752	1764	1775	1785	1795
1717	1729	1740	1753	1765	1776	1786	1796
1718	1730	1741	1754	1766	1777	1787	1797
1719	1731	1743	1755	1767	1778	1788	1798
1720	1732	1744	1756	1768	1779	1789	1799
1721	1733	1746	1757	1769	1780	1790	1800
1723	1734	1747	1758	1770	1781	1791	
1724	1735	1748	1759	1771			

On the London Bridge-Brighton line near Forest Hill is this 8-Car suburban train working a "roundabout" service

2-Car Motor Sets

(for South London and Wimbledon—West Croydon services)

1801	1805	1809
1802	1806	1810
1803	1807	1811
1804	1808	1812

MAIN LINE STOCK

2-Car Non-Corridor Motor Units

(Nol.)

1813	1829	1844*	1860
1814	1830	1845*	1861
1815	1831*	1846	1862
1816	1832	1847	1863
1817	1833	1848	1864
1818*	1834	1849	1865
1819	1835	1850	1866
1820	1836	1851	1867
1821	1837	1852	1868
1822	1838*	1853	1869
1823	1839*	1854	1870
1824	1840	1856	1871
1825	1841*	1857	1872
1826	1842	1858	1873
1827	1843*	1859	1874

2-car "Nols."—*continued*

1875	1879	1883	1887
1876	1880	1884	1888
1877	1881	1885	1889
1878	1882	1886	1890

* Without coupé compartment

2-car "Bils."—*continued*

2136	2141	2145	2149
2137	2142	2146	2150
2138	2143	2147	2151
2139	2144	2148	2152
2140			

† 88 3rd seats instead of 84

2-Car Motor Lavatory Sets

(Bil.)

2001†	2035	2068	2101
2002†	2036	2069	2103
2003†	2037	2070	2104
2004†	2038	2071	2105
2005†	2039	2072	2106
2006†	2040	2073	2107
2007†	2041	2074	2108
2008†	2042	2075	2109
2009†	2043	2076	2110
2010†	2044	2077	2111
2011	2045	2078	2112
2012	2046	2079	2113
2013	2047	2080	2114
2015	2048	2081	2115
2016	2049	2082	2116
2017	2050	2083	2117
2018	2051	2084	2118
2019	2052	2085	2119
2020	2053	2086	2120
2021	2054	2087	2121
2022	2055	2088	2122
2023	2056	2089	2123
2024	2057	2090	2124
2025	2058	2091	2125
2026	2059	2092	2126
2027	2060	2093	2127
2028	2061	2094	2128
2029	2062	2095	2129
2030	2063	2096	2130
2031	2064	2097	2132
2032	2065	2098	2133
2033	2066	2099	2134
2034	2067	2100	2135

2-Car Motor Lavatory Sets

(with one corridor and one non-corridor coach)

(Hal.)

2601	2624	2647	2670
2602	2625	2648	2671
2603	2626	2649	2672
2604	2627	2650	2673
2605	2628	2651	2674
2606	2629	2652	2675
2607	2630	2653	2676
2608	2631	2654	2677
2609	2632	2655	2678
2610	2633	2656	2679
2611	2634	2657	2680
2612	2635	2658	2681
2613	2636	2659	2682
2614	2637	2660	2683
2615	2638	2661	2684
2616	2639	2662	2685
2617	2640	2663	2686
2618	2641	2664	2687
2619	2642	2665	2688
2620	2643	2666	2689
2621	2644	2667	2690
2622	2645	2668	2691
2623	2646	2669	2692

4-Car Motor Lavatory Sets

(with three non-corridor and one 3rd/1st corridor coach)

(Lav.)

2921	2930	2939	2948
2922	2931	2940	2949
2923	2932	2941	2950
2924	2933	2942	2951
2925	2934	2943	2952
2926	2935	2944	2953
2927	2936	2945	2954
2928	2937	2946	2955
2929	2938	2947	

6-Car Motor Corridor Units

(with Pullman Car)

(Pul.)

3001	3006	3011	3016
3002	3007	3012	3017
3003	3008	3013	3018
3004	3009	3014	3019
3005	3010	3015	3020

6-Car Motor Corridor Units

(with Pantry Car)

(Pan.)

3021	3026	3030	3034
3022	3027	3031	3035
3023	3028	3032	3036
3024	3029	3033	3037
3025			

6-Car Motor Corridor Units

(with Pullman Car)
(For "City Limited" Service)

(Cit.)

3041 3042 3043

5-Car Pullman Motor Units

(For "Brighton Belle" Service)

(Bel.)

3051 3052* 3053

* This unit is temporarily withdrawn from service.

4-Car Kitchen Corridor Motor Units

(Res.)

3054	3061	3065	3069
3055	3062	3066	3070
3056	3063	3067	3071
3057	3064	3068	3072
3059			

4-Car Buffet Corridor Motor Units

(Buf.)

3073	3077	3080	3083
3074	3078	3081	3084
3075	3079	3082	3085
3076			

4-Car Corridor Motor Units

(Cor.)

3101	3114	3129	3143
3102	3115	3130	3145
3103	3116	3131	3146
3104	3118	3133	3147
3105	3120	3134	3148
3106	3121	3135	3149
3107	3122	3136	3150
3108	3123	3138	3151
3109	3124	3139	3152
3110	3125	3140	3153
3111	3126	3141	3154
3112	3127	3142	3155
3113	3128		

4-Car Suburban Sets

(SUB.)

4101	4109	4117	4125	4133	4141	4149	4216
4102	4110	4118	4126	4134	4142	4195	4217
4103	4111	4119	4127	4135	4143	4196	4218
4104	4112	4120	4128	4136	4144	4197	4219
4105	4113	4121	4129	4137	4145	4198	4220
4106	4114	4122	4130	4138	4146	4199	4221
4107	4115	4123	4131	4139	4147	4200	4250
4108	4116	4124	4132	4140	4148	4201	

N.B. Not all these units are yet completed

Part III

HEADCODES

Main Line

Western Section

(Including suburban routes, when worked by main line stock)

	SERVICE	ROUTE
2	Waterloo and Portsmouth Harbour	Chertsey
3	Waterloo and Portsmouth and Southsea	Bookham
4	Waterloo and Portsmouth Harbour	Bookham
5	Waterloo and Portsmouth and Southsea	Cobham
6	Waterloo and Portsmouth Harbour	Cobham
7	Waterloo and Portsmouth and Southsea	Worplesdon
8	Waterloo and Portsmouth Harbour	Worplesdon
9	Waterloo and Portsmouth and Southsea	Chertsey
10	Waterloo and Woking	Earlsfield
12	Waterloo and Alton	
13	Waterloo and Woking	Chertsey
14	Waterloo and Virginia Water	Weybridge
15	Waterloo and Alton	Chertsey
16	Waterloo and Guildford	Epsom
17	Waterloo and Windsor	Hounslow
	Staines and Weybridge	
	Waterloo and Horsham	Motspur Park
18	Waterloo and Windsor	Richmond
	Staines and Weybridge	
	Waterloo and Chessington South	Motspur Park
19	Waterloo and Epsom	Motspur Park
20	Waterloo and Ascot	Woking
24	Waterloo and Shepperton	Wimbledon
25	Waterloo and Woking	Brentford and Ascot
26	Waterloo and Woking	Richmond and Ascot
27	Waterloo and Reading	Brentford
28	Waterloo and Reading	Richmond
30	Waterloo and Hampton Court	
35	Portsmouth and Gillingham (Special)	Guildford and Epsom
42	Waterloo and Guildford	Cobham

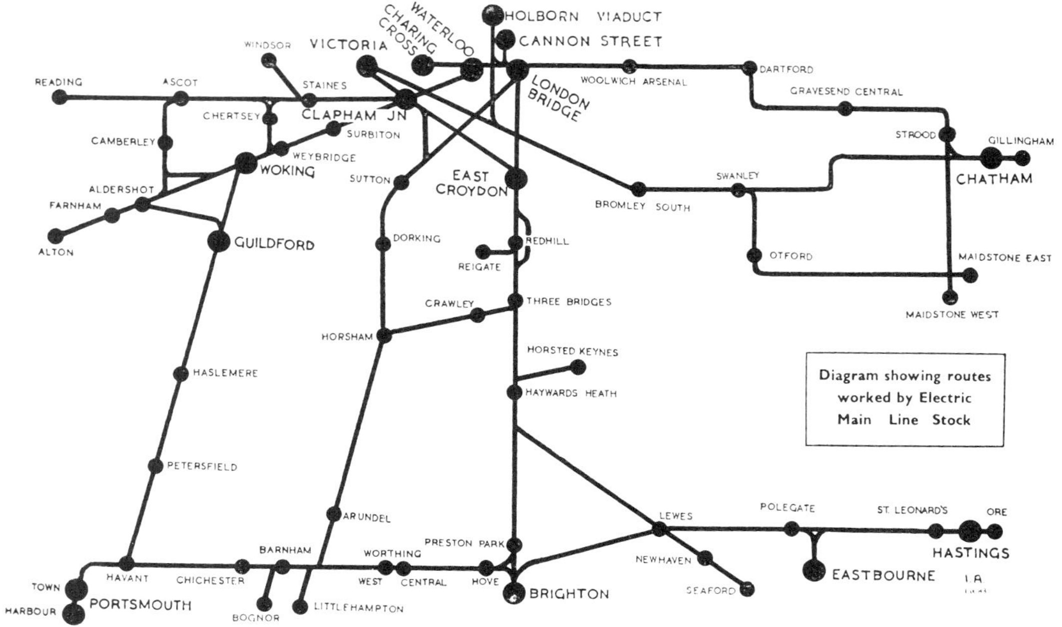
Diagram showing routes worked by Electric Main Line Stock
HOLBORN VIADUCT
CANNON STREET
WATERLOO
CHARING CROSS
VICTORIA
LONDON BRIDGE
WINDSOR
READING
ASCOT
STAINES
CLAPHAM JN
CHERTSEY
SURBITON
CAMBERLEY
WEYBRIDGE
WOKING
ALDERSHOT
FARNHAM
ALTON
GUILDFORD
SUTTON
EAST CROYDON
DORKING
REDHILL
REIGATE
CRAWLEY
THREE BRIDGES
HORSHAM
HORSTED KEYNES
HAYWARDS HEATH
HASLEMERE
PETERSFIELD
ARUNDEL
WOOLWICH ARSENAL
DARTFORD
GRAVESEND CENTRAL
STROOD
GILLINGHAM
CHATHAM
SWANLEY
BROMLEY SOUTH
OTFORD
MAIDSTONE EAST
MAIDSTONE WEST
PRESTON PARK
LEWES
POLEGATE
ST. LEONARD'S
ORE
HASTINGS
BARNHAM
WORTHING
WEST
CENTRAL
HOVE
NEWHAVEN
SEAFORD
EASTBOURNE
HAVANT
CHICHESTER
BRIGHTON
TOWN
HARBOUR
PORTSMOUTH
BOGNOR
LITTLEHAMPTON
I.A.

	SERVICE	ROUTE
43	Waterloo and Effingham Junction	Cobham
47	Waterloo and Shepperton	Richmond
61	Waterloo (M.L.) to Waterloo (W.L.)	Teddington
62	Waterloo (W.L.) to Waterloo (M.L.)	Teddington
80	Waterloo and Portsmouth Harbour (Non-Stop)	
86	Waterloo and Wimbledon	East Putney
87	Waterloo to Waterloo	Richmond–Hounslow
89	Waterloo to Waterloo	Hounslow–Richmond
03	Windsor and Portsmouth and Southsea	Worplesdon
04	Windsor and Portsmouth Harbour	Worplesdon
05	Waterloo and Portsmouth and Southsea	East Putney
06	Waterloo and Portsmouth Harbour	East Putney
07	Waterloo and Portsmouth and Southsea (Special)	Direct
08	Waterloo and Portsmouth Harbour (Special) ...	Direct

Empty trains provided with number headcodes bear a bar over relevant code

---

Central Section

N.B. The Quarry line is the fast road between Coulsdon and Earlswood, avoiding Redhill

1	Brighton and West Worthing	
	Arundel and Littlehampton	
	Eastbourne and Ore	
	Lewes and Seaford	
	Horsted Keynes and Haywards Heath	
2	Redhill and Reigate	
	Horsted Keynes and Brighton	
	Three Bridges and Horsham	
	Polegate and Eastbourne	
3	London Bridge and Brighton (Express)	Quarry
	Arundel and Bognor	Littlehampton
4	Victoria and Brighton (Express)	Quarry
	Arundel and Bognor	Direct
5	London Bridge and Brighton (Semi-fast)	Quarry
	Three Bridges and Littlehampton	Direct
6	Victoria and Brighton (Semi-fast)	Quarry
	Three Bridges and Bognor	Direct
7	London Bridge and Brighton (Slow)	Quarry
	Three Bridges and Bognor	Littlehampton
8	Victoria and Brighton (Slow)	Quarry
9	London Bridge and Brighton (Express)	Redhill
10	Victoria and Brighton (Express)	Redhill
12	Bognor and Portsmouth Harbour	
	Victoria and Brighton (Semi-fast)	Redhill
13	London Bridge and Brighton (Semi-fast)	Redhill
	Littlehampton and Portsmouth L.L.	
14	Victoria and Brighton (Slow)	Redhill
15	London Bridge and Brighton (Slow)	Redhill
	Chichester and Portsmouth L.L.	
16	Victoria and Littlehampton	Q. and Worthing
	Brighton to Ore	Eastbourne
17	London Bridge and Littlehampton	Q. and Worthing
	Brighton and Lewes	
18	Victoria and Littlehampton	Redhill and Worthing
	Ore to Brighton	Eastbourne
	Brighton to Eastbourne	
19	London Bridge and Littlehampton	Redhill, Worthing
20	Victoria and Portsmouth Harbour	Mitcham Junction
21	London Bridge and Portsmouth	Mitcham Junction
23	London Bridge and Portsmouth	Quarry and Horsham
24	Victoria and Brighton (Special)	Redhill
25	London Bridge and Portsmouth Harbour ...	Redhill and Horsham

	SERVICE	ROUTE
26	Victoria and Portsmouth Harbour	Quarry and Horsham
	Brighton and Hastings or Ore	Direct
27	London Bridge and Brighton (Special)	Quarry
28	Victoria and Portsmouth Harbour	Redhill and Horsham
	Brighton and Seaford	
29	London Bridge and Brighton (Special)	Redhill
30	Brighton and Portsmouth Harbour	Littlehampton
31	Charing Cross and Reigate	Redhill
	Brighton and Bognor Regis	Littlehampton
32	Brighton and West Worthing	Preston Park
34	Victoria and Reigate	Redhill
35	Brighton and Littlehampton	
	London Bridge and Portsmouth and Southsea or Portsmouth Harbour	W. Croydon & Effingham Junction
37	London Bridge and Reigate	Redhill
	Horsted Keynes, Haywards Heath and Seaford	
38	Streatham Hill and Brighton	Quarry
	Tooting and Littlehampton	Wimbledon and Mitcham
39	London Bridge and Brighton	Streatham and Quarry
40	Victoria and Bognor Regis	Mitcham Junction
41	London Bridge and Bognor Regis	Mitcham Junction
42	Victoria and Brighton (Special)	Quarry
43	London Bridge and Bognor Regis	Quarry and Horsham
45	London Bridge and Bognor Regis	Redhill and Horsham
46	Victoria and Bognor Regis	Quarry and Horsham
	Horsted Keynes, Haywards Heath and Lewes	
47	London Bridge and Bognor Regis	West Croydon
48	Victoria and Bognor Regis	Redhill and Horsham
49	Horsted Keynes, Haywards Heath and Eastbourne	
	Tooting and Bognor Regis (not via Littlehampton)	Wimbledon and Mitcham
50	Victoria and Littlehampton	Mitcham Junction and Horsham
51	London Bridge and Littlehampton	Mitcham Junction and Horsham
52	Victoria and Ore	Quarry and Eastbourne
53	London Bridge and Littlehampton	Quarry and Horsham
54	Victoria and Ore (not via Eastbourne)	Quarry and Direct
56	Victoria and Littlehampton	Quarry and Horsham
57	London Bridge and Littlehampton	West Croydon and Horsham
58	Victoria and Littlehampton	Redhill and Horsham
59	London Bridge and Littlehampton	Redhill and Horsham
60	Victoria and Ore (not via Eastbourne) (Semi-fast)	Redhill and Direct
	Brighton and Portsmouth Harbour	Direct
61	London Bridge and Ore	Redhill and Eastbourne
62	Victoria and Eastbourne	Quarry
	Brighton and Portsmouth Harbour (Slow) ...	Direct
63	London Bridge and Ore	Quarry and Eastbourne
64	Victoria and Eastbourne	Redhill
	Brighton and Bognor Regis	Direct
65	London Bridge and Eastbourne	Quarry
67	London Bridge and Eastbourne	Redhill
68	Victoria and Seaford	Quarry
69	London Bridge and Seaford	Quarry
70	Victoria and Portsmouth Harbour...	Mitcham Junction and Littlehampton
71	London Bridge and Portsmouth Harbour ...	Mitcham Junction and Littlehampton
72	Victoria and Ore	Redhill and Eastbourne
73	London Bridge and Portsmouth Harbour ...	Quarry, Horsham and Littlehampton
74	Victoria and Seaford	Redhill

	SERVICE	ROUTE
75	London Bridge and Portsmouth Harbour ...	Redhill, Horsham and Littlehampton
76	Victoria and Portsmouth Harbour	Quarry, Horsham and Littlehampton
78	Victoria and Portsmouth Harbour	Redhill, Horsham and Littlehampton
79	London Bridge and Seaford	Redhill
90	Victoria and Bognor Regis	Mitcham Junction and Littlehampton
91	London Bridge and Bognor Regis	Mitcham Junction and Littlehampton
92	Victoria and Bognor Regis	West Croydon and Littlehampton
93	London Bridge and Bognor Regis	Quarry, Horsham and Littlehampton
95	London Bridge and Bognor Regis	Redhill, Horsham and Littlehampton
96	Victoria and Bognor Regis	Quarry, Horsham and Littlehampton
98	Victoria and Bognor Regis	Redhill, Horsham and Littlehampton

Eastern Section

42	Charing Cross and Gillingham	Parks Bridge & Loop Line
43	Cannon Street and Gillingham	Parks Bridge & Loop Line
46	Charing Cross and Maidstone West	Parks Bridge & Loop Line
47	Cannon Street and Maidstone West	Parks Bridge & Loop Line
52	Charing Cross and Gillingham	Lewisham and Loop Line
53	Cannon Street and Gillingham	Lewisham and Loop Line
56	Charing Cross and Maidstone West	Lewisham and Loop Line
57	Cannon Street and Maidstone West	Lewisham and Loop Line
62	Charing Cross and Gillingham	North Kent Line
63	Cannon Street and Gillingham	North Kent Line.
64	Charing Cross and Maidstone West	North Kent Line
65	Cannon Street and Maidstone West	North Kent Line
72	Charing Cross and Gillingham	Bexleyheath
73	Cannon Street and Gillingham	Bexleyheath
74	Charing Cross and Maidstone West	Bexleyheath
75	Cannon Street and Maidstone West	Bexleyheath
82	Charing Cross and Gillingham	Greenwich
83	Cannon Street and Gillingham	Greenwich
84	Charing Cross and Maidstone West	Greenwich
85	Cannon Street and Maidstone West	Greenwich
90	Victoria and Gillingham	Herne Hill
91	Holborn Viaduct and Gillingham	Herne Hill
92	Victoria and Gillingham	Catford Loop
93	Holborn Viaduct and Gillingham	Catford Loop
94	Victoria and Maidstone East	Herne Hill
95	Holborn Viaduct and Maidstone East	Herne Hill
96	Victoria and Maidstone East	Catford Loop
97	Holborn Viaduct and Maidstone East	Catford Loop

A Portsmouth main line motor coach

HEADCODES

Suburban Services

Western Section

	SERVICE	ROUTE
H	Waterloo and Hampton Court	
	Waterloo and Windsor	Hounslow
$\overline{H}$	Waterloo and Guildford	Cobham
	Waterloo and Windsor	Richmond
I	Waterloo and Dorking North	Motspur Park
	Waterloo and Reading	Brentford
$\overline{I}$	Waterloo and Effingham Junction	Motspur Park
L	Waterloo and Chessington South	
$\overline{L}$	Waterloo and Leatherhead *	Motspur Park
O	Waterloo (W.L.) to Waterloo (W.L.)	Hounslow and Richmond
$\overline{O}$	Waterloo (W.L.) to Waterloo (W.L.)	Richmond and Hounslow
P	Waterloo and Wimbledon	East Putney
$\overline{P}$	Waterloo and Wimbledon Park	East Putney
S	Waterloo and Shepperton	Kingston
$\overline{S}$	Waterloo and Shepperton	Richmond
T	Waterloo and Woking	
$\overline{T}$	Waterloo and Chertsey	Weybridge
V	Waterloo (M.L.) to Waterloo (W.L.)	Teddington
$\overline{V}$	Waterloo (W.L.) to Waterloo (M.L.)	Teddington
Λ	Waterloo and Strawberry Hill	Kingston
$\overline{\Lambda}$	Waterloo and Strawberry Hill	Richmond

---

Central Section

H	London Bridge to Victoria	Tulse Hill and Streatham Hill
$\dot{H}$	Victoria to London Bridge	Streatham Hill and Tulse Hill
I	Victoria and Coulsdon North	Streatham Common
$\overline{I}$	Victoria and Tattenham Corner or Caterham ...	E. Croydon direct
$\dot{I}$	London Bridge and Coulsdon North	Forest Hill
$\ddot{I}$	London Bridge and Coulsdon North	Tulse Hill and Crystal Palace
$\dot{L}$	London Bridge and Dorking North	Mitcham Junction
$\ddot{L}$	London Bridge and Effingham Junction	Mitcham Junction
$\dot{O}$	London Bridge and Epsom Downs	Norwood Junction and West Croydon
$\ddot{O}$	London Bridge and Wallington	Ditto

	SERVICE	ROUTE
P̄	Holborn Viaduct and West Croydon	Tulse Hill and St. Helier
	Victoria and West Croydon	Streatham Hill
Ṗ	London Bridge to London Bridge	Sydenham, Crystal Palace (L.L.) and Tulse Hill
P̈	London Bridge to London Bridge	Tulse Hill, Crystal Palace (L.L.) and Sydenham
S	Victoria and Wallington or Sutton	Streatham Common
S̄	Victoria and Selhurst or West Croydon	Streatham Common
Ṡ	Victoria and Epsom Downs	Streatham Common
	Charing Cross or Cannon Street and Caterham	East Croydon direct
	Tattenham Corner or Caterham to Cannon Street	East Croydon direct
S̈	Charing Cross or Cannon Street to Tattenham Corner	East Croydon direct
	Tattenham Corner or Caterham to Charing Cross...	East Croydon direct
T̈	London Bridge and Tattenham Corner or Caterham	East Croydon direct
V	Victoria and Epsom or Guildford	Mitcham Jct. & Bookham
V̇	London Bridge to London Bridge	Norwood Jct., Thornton Heath and Tulse Hill
V̈	London Bridge to London Bridge	Tulse Hill, Thornton Heath & Norwood Jct.
1̄	East Croydon and Victoria	Direct
2	London Bridge and Victoria	South London Line
2	West Croydon and Wimbledon	Mitcham

Eastern Section

SERVICES FROM CHARING CROSS AND CANNON STREET

Main Line (Orpington, Sevenoaks, etc.)

	SERVICE	ROUTE
P̈	London to Bromley North	Grove Park
	Bromley North to Charing Cross	Grove Park
Ṗ	Bromley North to Cannon Street	Grove Park
O	London and Orpington	Grove Park
Λ	London and Sevenoaks	Grove Park

Except service **P** *above, headcode modification for* :—

Down Trains	*serving Lewisham ... ...*	*Two dots*
Up Trains	*to Charing Cross via Park's Bridge ...*	*A Bar*
	to Charing Cross serving Lewisham ...	*Two dots*
	to Cannon Street via Park's Bridge ...	*No modification*
	to Cannon Street serving Lewisham ...	*One dot*

North Kent Line (to Dartford, etc.)

	SERVICE	ROUTE
L	London to Dartford	Sidcup
	Gillingham line to Cannon Street	Sidcup
L̄	London to Gillingham line	Sidcup
	Gillingham line to Charing Cross	Sidcup
L̇	London to Dartford	Lewisham, Sidcup
	Gillingham line to Cannon Street	Lewisham, Sidcup
L̈	London to Gillingham line	Lewisham, Sidcup
	Gillingham line to Charing Cross	Lewisham, Sidcup

	SERVICE	ROUTE
P	London and Dartford or Gillingham line	Blackheath and Charlton
S	London and Dartford or Gillingham line	Bexleyheath
V	London and Dartford or Gillingham line	Greenwich

Except service **L** *:—Headcodes for Dartford trains are unmodified. Gillingham and Gravesend headcodes bear a bar thereover.*

Up Trains *to Cannon Street*	*Headcode unmodified*
to Charing Cross	*A Bar thereover*

Mid-Kent Line (to Sanderstead, etc.)

H	London and Hayes	Lewisham
I	London and Beckenham Junction	Lewisham
$\dot{I}$	London to Sanderstead	Lewisham
$\ddot{I}$	London to Sanderstead	Ladywell Loop
T	London and Addiscombe	Lewisham

Except services $\dot{I}$, $\ddot{I}$ *on* **Down** *journey headcode modification for:—*

Down Trains *via Ladywell loop*	*A Bar*
All **Up Trains** *to Charing Cross via Lewisham* ...	*A Bar*
via Ladywell Loop	*Two Dots*
to Cannon Street via Lewisham	*No modification*
via Ladywell Loop	*A Dot*

SERVICES FROM VICTORIA AND HOLBORN VIADUCT

H	London and Herne Hill	
H	Orpington to Holborn Viaduct	Herne Hill
H	Sevenoaks to Holborn Viaduct	Herne Hill
$\dot{H}$	Sevenoaks to Holborn Viaduct	Nunhead
$\ddot{H}$	Crystal Palace to Holborn Viaduct	Nunhead
$\bar{H}$	Orpington to Holborn Viaduct	Nunhead
J	Holborn Viaduct and Blackfriars to Dartford ...	London Bridge or Nunhead and Bexleyheath
$\dot{O}$	Victoria to Orpington	Herne Hill
$\ddot{O}$	Holborn Viaduct to Orpington	Herne Hill
$\bar{O}$	London to Orpington	Nunhead
S	London to Sevenoaks	Herne Hill
S	Orpington to Blackfriars	Herne Hill
S	Sevenoaks to Blackfriars	Herne Hill
$\dot{S}$	Sevenoaks to Holborn Viaduct	Nunhead
$\ddot{S}$	Crystal Palace to Blackfriars	Nunhead
$\bar{S}$	London to Sevenoaks	Nunhead
U	Holborn Viaduct or Blackfriars and Dartford Line	London Bridge, Sidcup
$\dot{U}$	Holborn Viaduct or Blackfriars and Dartford Line	Nunhead, Sidcup
V	Orpington to Victoria	Herne Hill
$\dot{V}$	Sevenoaks to Victoria	Herne Hill